William Pierson Judson

City Roads and Pavements Suited to Oswego, New York

William Pierson Judson

City Roads and Pavements Suited to Oswego, New York

ISBN/EAN: 9783744678308

Printed in Europe, USA, Canada, Australia, Japan

Cover: Foto ©berggeist007 / pixelio.de

More available books at **www.hansebooks.com**

CITY

ROADS AND PAVEMENTS

SUITED TO OSWEGO, NEW YORK.

BY

WILLIAM PIERSON JUDSON,

M. Am Soc. C. E.,   M. Inst. C. E.

OSWEGO, N. Y.:
R. J. OLIPHANT, PRINTER AND BOOKBINDER.
1884.

PREFACE.

The subject of pavements and roads is here presented at the request of Mayor John D. Higgins, for the consideration and discussion of the Oswego common council and board of public works, and of the citizens who may be interested.

If this presentation were made for the usual purpose and in the usual form of a consulting engineer's project for an actual work, there could then be briefly stated the best methods for the several parts, with estimates of their cost and with specifications for their detailed execution.

The present purpose, however, is not the making of pavement, but the forming of public opinion. For this it is necessary to describe the various methods which may be and will be proposed, and the several reasons why some are to be avoided and some are to be accepted and used.

The application is therefore not limited to Oswego alone, but may equally well reach almost any city containing twenty to thirty thousand people who are awakening to the need and the gain of having good roads.

Such cities should profit by the costly experiences of others, so far as they can be applied to the case, and thus avoid building apparently cheap roads whose repairs will afterward make them cost more than the most costly, or costly ones which are not adapted to the local needs or means.

The general reports and results of seventy-five cities, together with many recently published articles by experienced engineers, have, with personal experience, formed the basis of this discussion; while most of the data relating to brick pavements has just been obtained by direct application to the city engineers of various cities, and by personal examination of pavements in progress and in use.

W. P. J.

Oswego, New York,
October 1, 1894.

CONTENTS.

CHAPTER I.

LOCAL CONDITIONS.

CHAPTER II.

THE VARIOUS PAVEMENTS.

CHAPTER III.

WOOD-BLOCK PAVEMENTS.

CHAPTER IV.

BLOCK-STONE PAVEMENTS.

CHAPTER V.

ASPHALT PAVEMENTS.

CHAPTER VI.

BROKEN-STONE ROADS.

CHAPTER IX.

CONCLUSIONS—LOCAL APPLICATIONS TO OSWEGO.

.

ILLUSTRATIONS.

CITY ROADS AND PAVEMENTS

SUITED TO OSWEGO, NEW YORK.

CHAPTER I.

LOCAL CONDITIONS.

OSWEGO'S STREET AREA.

The extent of Oswego's street surface is enormous in proportion to the population, the traffic, and the treasury. The city map shows 53 per cent of the area of the improved portion to be streets, of which about one-half is actual roadway.

Of these roadways, there are shown upon the map within the city limits over 123 miles, of which about 74 miles are actually laid out and more or less built upon and improved. This total is approximately made up of the following items:

	MILES.
Cobble-stone pavement...............................	0.95
Cobble-stone pavement with recently built central roadway of Oswego sandstone blocks................	1.60
Oswego sandstone block pavement.....................	0.37
Potsdam or Hammond sandstone block pavement, recently built in place of a cedar block pavement....	0.12
North River bluestone...............................	0.06
Macadam and Telford roads, made of crushed fieldstone and crushed limestone...........................	0.90
Bank gravel and loam road...........................	0.60

Dirt roads with more or less surfacing of crushed field-
stone 4.00
Dirt roads surfaced with lake shore gravel and bank
gravel 2.40

Total improved roads...................... 11.00
Dirt roads with natural earth roadbed............... 63.00

Sum total................................. 74.00

None of these roads or pavements are sub-drained,
the natural slopes of the city site favoring surface drainage on many of the streets. On many other streets surface water lies until it evaporates. To improve any considerable part of this at once is impossible. To perfect
enough to serve as an object lesson and an incentive, can
be accomplished within a year.

REDUCTION OF SURFACE TO BE PAVED.

Let it be seen if the total area of roadways cannot
be materially reduced.

When the length and present width of the roadways
are computed, the least cost per square yard of doing
anything whatever to them makes an appalling total.

The 74 miles of length must be accepted. The width
of roadway of the business streets, 54 feet, is not too
great. But the roadway of the residence streets—nine-
tenths of the whole—can be reduced one-fourth in width.

The streets are 100 feet wide between the house
fronts, or 66 feet between fence lines, with six feet for sidewalks and six feet for bermes on each side of the roadway.
These roadways are now 42 feet wide, ample for five
carriages abreast. Has anyone ever seen five carriages
abreast on the residence streets? and under what cir-
cumstances can such an array ever be desirable? Or has

anyone ever seen one-half of a 42-foot roadway in actual use? On the other hand, is there any resident who has not found some of the present sidewalks too narrow? or who has not been crowded off from the sidewalk and among the trees, when meeting an outcoming school or a church congregation?

If the width of the bermes on each side of the street is increased to 11 feet, this will reduce the roadway to 32 feet, which is the standard width most often used in other cities, and will also add to the beauty of the streets by a wider grassy berme outside the row of trees. This arrangement has existed for many years on West Fifth street facing the park. No complaints have ever been heard that the roadway at that point was too narrow. The same arrangement has just been made with good effect on West Third street south of Bridge street.

This wider berme will also permit an extra line of sidewalk outside the row of trees near churches and schools where the present walks are too narrow. The loose earth and ashes which must be scraped from the present roadways before attempting to form new ones, will supply on the spot the material to form the wider bermes.

This will reduce the area of roadway one-fourth.

DRAINAGE OF ROADWAYS.

The thorough drainage of such streets as are now naturally muddy in spring and fall must be provided for before any method of paving or surfacing is considered.

The natural earth is the real roadbed which does the work, and it can only support the pavement—of whatsoever kind it may be—by being kept dry.

The Oswego streets which run east and west have

generally good grades, and will drain naturally toward the river in most cases, if rightly formed. But most of the numbered streets, running north and south, on which rainwater stands until it evaporates, or which have springs in the subsoil, must have sub-drainage by tile drains before any form of surface or of pavement will be of any permanent value or effect.

The amount of water which falls upon these wide streets, and which must lie upon the surface until it evaporates, is very great. For instance, West Third street from Bridge street to the lake is one-half mile in length. Upon this the direct rainfall between fence lines is equal to 16,000 tons, or 4,000,000 gallons of water per year, for which no drainage whatever is provided.

Any such roadbed, where, from any cause, water naturally stands and forms mud, must be thoroughly sub-drained. To put broken stone, or gravel, or any valuable material of any kind upon a bed of earth and ashes which rain will convert to mud, is to throw money and material away.

One line, or, in some cases, two lines of three-inch or four-inch porous tile, laid lengthwise of the road, away from the reach of the shade-tree roots, below frost, and leading to sewer manholes, must be the first move toward making any permanent roadway on a flat street.

ROLLING THE EARTH ROADBED.

For any method of road-making or of paving which may be adopted, a steam roller of about twelve tons weight is requisite in order to compact the earth roadbed so that it will sustain the wheels which will pass over it.

As well try to make the bricks of old Egypt without straw as to try to make the roads of to-day without a heavy steam roller. Every fully equipped road-builder has one or more.

Oswego has never made any effective effort to have good roads, as is shown by comparison with other cities of similar size, and no good results can be expected until the proper tools are used. For any system of pavements or roads a heavy roller is the thing first needed, and no contractor's bid should be considered unless he agrees to use one. Those who question this have only to examine the recently built block-stone pavement at and near the intersection of West Second and Bridge streets. The undulations and hollows in the surface of the pavement result in part from the lack of a proper roller, which would first have disclosed the soft places in the earth roadbed, and then would have packed the grading material into them, so that the finished pavement would have had a solid and regular foundation and a permanent surface.

GOOD EFFECT OF ROLLER ON DIRT ROADS.

Especially valuable would such a roller be for Oswego's great extent of dirt roads, which could be formed by use of the wheeled scraper (now standing idle), and then rolled to a smooth, hard surface, thus furnishing fine roadways during the summer months until the fall rains make them muddy.

Nearly a mile per day of such temporary roadway could be made at small cost by a scraper, sprinkler, and steam roller working together. The scraper and sprinkler the city now has; the roller may cost $4,000 to purchase and $6 per day to operate. Its selection should include the points of being not too heavy to safely cross the bridges, and of having power enough to climb 10 per cent grades when they are covered with loose road material, and also of having a record for durability under rough usage.

WIDE TIRES ON WHEELS.

To supplement the good effect of a roller on the dirt roads, which are now cut by narrow tires, the use of wide tires on heavy wagons should be required. The following is a practicable way of initiating such a rule:

Let the Board of Public Works order that no wagon will be employed upon city work unless it has five-inch tires on rear wheels and four-inch tires on front wheels, with the front axle eight inches shorter than the rear axle. This will make each wagon equal to two nine-inch rollers.

Let the same order be applied to the street-sprinkling wagons and to public carters, as a condition of issuing a license. A future date could be published at which all heavy wagons doing business in the city shall have such wheels, when the roadways will no longer be so deeply cut and furrowed as now by the pressure of traffic.

PRESSURE OF TRAFFIC.

It is only necessary to consider the great pressure which ordinary traffic will put upon the roadbed in order to realize that no pavement can keep its form and its regular surface unless the earth roadbed, on which all the pressure finally comes, has been perfectly compacted before the pavement is laid over it; for the pavement, of whatever material it may be, is merely a more or less rigid surface which receives the pressure of traffic and distributes it to the supporting earth. For instance, the ordinary coal wagon, weighing 1,200 pounds, draws two tons of coal and has tires two inches wide. As the wagon stands on the pavement, the bearing surface does not exceed a length of one and one-half inches on each wheel; the four wheels thus standing upon a total surface of twelve square inches, with a total pressure of 5,200 pounds, or 433 pounds per square inch, and this is applied with a rolling pressure which is most destructive.

COMPARISON WITH PRESSURE OF STRUCTURES.

The degree of pressure which this puts upon any pavement will be best appreciated by comparing it with the pressures per square inch upon the clay, sand, or earth underlying the foundations of some well-known great structures.

The Cleveland viaduct............... 14 to 23 lbs. per sq. in.
The 1894 London tower bridge............. 21   "   "
The sixteen-story office buildings of Chicago.. 21   "   "
The Memphis bridge piers................. 22   "   "
The Albany capitol....................... 28   "   "
The Brooklyn bridge anchorage........... 56   "   "

The earth supporting these structures is, of course, compressed to the greatest degree in its natural formation, but the average pressure of these structures is less than one-sixteenth of the pressure concentrated on an ordinary wagon wheel.

Chapter II.

THE VARIOUS PAVEMENTS.

Having indicated these preliminaries for reducing the area of the roadways, for draining them, and for rolling them, there may now be considered the various methods by which the roadways may be improved, taking first those methods which have proved most costly, and later those which are most likely to suit the local needs and treasury.

COMPARATIVE LOADS ON VARIOUS PAVEMENTS.

In considering the desirability of the different road surfaces and pavements, it may be noted that a team drawing one ton on a good dirt road can with the same effort take two tons over a good macadam surface. Passing from this to a good block-stone pavement, six tons could be drawn as easily, and this load can be increased to eight tons on wood-block or vitrified brick, or to ten tons on an asphalt pavement.

COST OF PAVEMENTS ELSEWHERE.

The following conclusions regarding wood, block-stone, and sheet asphalt pavements are based upon the experiences of the 32 following-named cities, eight of which have wood-block pavements, 27 of which have sheet asphalt pavements, and all of which have block-stone, six being sandstone and the rest granite.

CITY AND STATE.	BLOCK-STONE.		SHEET ASPHALT.		WOOD.	
	Granite.	Sandstone.			Cedar-block.	
	Cost. Sq. Yard.	Cost. Sq. Yard.	Miles.	Cost. Sq. Yard.	Miles.	Least Cost.
Albany, N. Y.	$2.90			$3.12		
Allegheny, Pa.	3.37			2.75		
Atlanta, Ga.	1.49			3.00		
Boston, Mass.	3.90		4	3.30		
Brooklyn, N. Y.	2.33		11	3.00		
Buffalo, N. Y.		$3.25	150	3.50		
Chicago, Ill.	3.00		24	2.90	648	$1.10
Cincinnati, Ohio.	4.20			3.00		
Columbus, Ohio	3.71		11	2.54		
Denver, Col.	3.40		4	2.35		
Detroit, Mich.	4.25			3.20		
Kansas City, Kan.	2.74		2	2.55	26	1.50
Kansas City, Mo.		2.90	16	2.80	43	1.35
Milwaukee, Wis.	2.37			2.93	47	1.05
Minneapolis, Minn.	1.67		2	2.75	63	.76
Nashville, Tenn.	2.40					
New Orleans, La.	4.75		8	3.65		
New York, N. Y.	3.50		52	3.00		
Omaha, Neb.	2.32		23	2.68	38	1.52
Oswego, N. Y.		2.45				
Philadelphia, Pa.	2.41			2.50		
Pittsburg, Pa.	2.38			3.35		
Portland, Me.	2.00					
Providence, R. I.	3.25			2.65		
Rochester, N. Y.		1.90 / 3.00	9	2.60		
San Francisco, Cal.	2.00					
St. Paul, Minn.	2.05			2.70	30	1.10
Syracuse, N. Y.	1.15	3.00	4½	2.45		
Toledo, Ohio.	3.56		10	2.50		
Utica, N. Y.	3.20	2.50		1.95½		
Washington, D. C.	3.15		125	2.25		
Wilmington, Del.	2.08					
Average of prices.	$2.90	$2.71		$2.81		$1.19

PAVEMENTS FOR STEEP GRADES.

In selecting a pavement for a given street, the choice will often be limited by the fact that the grade is too steep to permit the use of a pavement which would otherwise be preferred.

The most useful information upon this subject can be obtained from the teamsters and horsemen of cities in which different pavements on varying grades have been in use. If it is generally agreed that certain pavements are shunned by teamsters because their horses slip and fall when going down a certain street with a load, it will evidently be unwise to repeat the construction of a similar pavement with the same slope elsewhere.

Asphalt Pavements.—The practical limit of slope on business streets is 1 foot in 25, or 4 per cent, though any slope steeper than 1 in 33, or 3 per cent, is not advisable on a main thoroughfare. On residence streets it has in many cases been laid on grades as steep as 6 per cent, and in some cases $7\frac{1}{2}$ per cent, or 1 in $13\frac{1}{2}$; the residents accepting the inconveniences resulting from a few days of icy roadway because of the advantages during the rest of the year. On semi-business streets with a grade as steep as 6 per cent, a good arrangement is to lay a 16-foot asphalt roadway in the center, with an 8-foot strip of block-stone with cement joints on each side. Even on the flat streets, however, in cold, misty weather horses slip badly, so that in Washington it is common to remove the shoes from horses in winter because the bare hoofs slip less.

Vitrified Brick.—No complaints are made of slipping upon grades of 1 in 20, or 5 per cent. But there will be more or less slipping as soon as this slope is exceeded. Upon brick pavements varying from level streets up to 1 foot in 9, or 11 per cent, it is found that horses slip only upon those having a grade of 1 in $16\frac{2}{3}$, or 6 per cent, and over, so that for any slope over 5 per cent it will be advisable to use special brick having a beveled or rounded top. With this precaution, there is no reason why vitrified brick should not be used upon slopes of 1 to $8\frac{1}{3}$, or 12 per cent.

Cedar Blocks.—The same conditions should be applied to these as to brick.

Block Stone.—This may be used in its ordinary form upon slopes less than 1 in 10, but for this slope and greater the blocks should have chamfered tops to give better foothold.

This general subject is fully discussed by Messrs. Rudolph Hering and Andrew Rosewater, members of the American Society of Civil Engineers, in the *Engineering News* of February 19, 1891.

FALLS ON DIFFERENT PAVEMENTS.

As to the relative liability to accidents from slipping of horses' feet upon different pavements, observations were made for Captain F. V. Greene during a period of six months in ten different cities, when over 800,000 horses were observed, with the result of showing that a horse may travel, for each fall that occurs—

272 miles on wood-block pavement.
413 "   " granite-block   "
583 "   " sheet-asphalt   "

These results differ radically from those obtained by Colonel Haywood in London, where it is required that horses shall be smooth-shod instead of having sharp toe-calks, which are generally used in the United States, and where rock asphalt is used instead of Trinidad asphalt. The results observed in London were—

446 miles on wood-block pavement.*
132 "   " granite-block   "
191 "   " sheet-asphalt   "

CAR TRACK CONSTRUCTION.

When any of these pavements are used upon a street containing car tracks, special attention must be given to the construction of the track and of the pavement next the rails, or the pavement will be heaved and broken by the yielding of the rails under the weight of passing cars.

The railroad company should be compelled to use heavy rails with a grooved top, and to make the track structure as nearly rigid as possible.

On each side of each rail should be set a four-inch line of granite or sandstone blocks as deep as the entire pavement and its foundation, and standing upon a line of concrete base four or more inches below the concrete of the pavement. The tops of these stone blocks should be set exactly even with the tops of the rails, which will not be likely to heave them.

Chapter III.

WOOD–BLOCK PAVEMENTS.

USE IN THE UNITED STATES.

Fifteen to twenty years ago several cheap forms of wood-block pavements were in general favor and wide use throughout the United States. There were few cities which did not try the Nicholson block or some other form, but the pavement was soon generally discredited by bad construction and worse maintenance, and was generally displaced, though a few western cities having a cheap wood supply and an extended street surface are still using it largely.

The usual construction is to lay a foundation of plank one and one-half or two or three inches thick, covered with three inches of sand, upon which are placed, on end, round seven-inch white cedar blocks of three to ten inches diameter; the irregular spaces between these blocks being filled with a bituminous concrete formed of coal-tar and gravel.

This construction is a cheap one adapted to the needs of a rapidly growing city, and stands well for two or three years, and is then noiseless and easy for traction. It then begins to absorb filth and to give off bad odors, and to become more and more rough and irregular. The blocks wear so differently that repairs are costly, and, after an average life of seven years, it must be wholly renewed.

Chicago now has 648 miles of this pavement—more than the combined mileage of all other cities which report its use. The superintendent of streets writes that its repairs cost three times as much as the repairs of other pavements.

The original cost varies from the minimum of 75⁷⁄₁₀ cents per square yard on plank at Minneapolis, to 85 cents per square yard at Toledo (whose mayor writes that other pavements would in the end cost less), to $1.10 on plank at Chicago, $1.45 on concrete at St. Paul, and $1.55 on concrete at Kansas City, Mo.; the average of the lowest cost to these eight cities now using it being $1.19 per square yard.

RECENT FOREIGN WOOD PAVEMENTS.

In trying to find some relief from the objectionable features of block-stone pavements, which have only been endured because there seemed to be no satisfactory substitute, the European and Australian cities are, during recent years, using various forms of wood-block pavements, just as the cities of the United States have been turning their attention to pavements of vitrified brick, the use of which on a large scale is confined to this country, though Holland has thus used it more or less for a century.

The foreign wood-block pavements are of much better construction and of higher cost than those above described as in use in the cities of the United States. London has given special attention to various forms of improved wood-block pavements, all of which have a solid and firm foundation, usually of concrete, upon which blocks are set which are of seasoned wood, free from sap and usually treated with creosote oil or some preservative. The experience has been in London that such pavements, with necessary repairs, will endure heavy traffic for ten years; the original cost being $3.90 to $4.38 per square yard, and the annual repairs costing 36 to 64 cents per square yard. The most recent method adopted in London during 1894 is to put a wrought-iron band on two sides of each block, the slightly projecting edge giving a foothold for horses and reducing the rate of wear of the wood.

Sidney, Australia, has twelve miles of wood-paved streets upon which Australian hardwoods have been used with most remarkable results, which would be incredible if not well substantiated by the statements, under date of June, 1894, of Mr. W. A. Smith, M. Inst. C. E., who is the engineer in charge. Queen street, which has an estimated daily traffic of 25,000 tons, has been thus paved, and its blocks after eight years' use show a greatest observable wear of one-sixteenth of an inch, and are otherwise in almost as good a state as when laid. The original cost was $1.35 per square yard exclusive of foundations, with an annual cost of ten cents per square yard for maintenance. The foundation of the Sidney pavement was a nine-inch layer of concrete. On this the hardwood paving blocks of turpentine wood, spotted gum, black-butt, and tallowwood, six inches deep, were laid, after having been twice dipped in boiling tar and stacked two days to allow the surplus tar to drain off. On grades as steep as 1 in 20 the joints were made one-fourth inch wide and were caulked with tar-pitch and basalt screenings. Where the joints had been filled with hydraulic cement, the wood blocks of all species were found to be destroyed by dry rot, but with the construction described the pavements are free from the

various faults of our cedar blocks and are expected to
have a life of 21 years, equaling asphalt, which could
not be used on the same grades. In Melbourne similar
pavement is estimated to last 14 years. Either of these
improved methods, or the more crude ones generally
used in this country, are costly; the final expense of our
cheap construction being twice as great as for asphalt or
for granite blocks, and probably much greater than if
white oak or some similar hardwood were used.

Chapter IV.

BLOCK-STONE PAVEMENTS.

LOCAL USE.

The only pavement which Oswego now has, except the old cobble-stone pavements, is of Potsdam sandstone blocks on gravel and sand foundation with sand joints. The most recent section is that on West Bridge street from First to Third streets, which cost $2.45 per square yard for the lower part and $2.40 for the extension now building to Third street.

A similar but very much better pavement was recently built in Rochester upon a solidly rolled bed and a broken stone foundation, the paving being of Medina sandstone blocks, whose surface was very regular and whose joints were filled with hydraulic cement. The finished cost was $1.90 per square yard.

RECOGNIZED EVILS OF BLOCK-STONE PAVEMENTS.

In examining the records of the great cities which do the best street work and employ the best skill to plan and execute it, it appears that block-stone pavement of all kinds has long been regarded as a necessary evil which has only been tolerated because it was an improvement on the original earth and on the barbarous cobble-stone pavement which formed the first stepping-stone out of the mud.

Block-stone pavements are only properly used for those parts of the streets of large cities where the heaviest traffic demands it. There is no such excessively

heavy traffic in Oswego, nor in any city of its size.
For such traffic the best granite or trap, in small blocks,
laid on a concrete foundation, is necessary.

If the traffic is heavy enough to demand block-stone
at all, the sandstone blocks will be worn out in five or
six years and then need renewal.

If the traffic is light enough, as in Oswego, for sandstone to stand for many years without much wear, as it
does in Oswego, then block-stone of any sort is needless.
It is costly, and noisy, and filthy, especially if laid with
open joints, the joints holding dirt and dust and becom-
ing worse with wear. Where it is not a necessity, other
pavements are better in every way. Either sheet asphalt
or vitrified brick have repeatedly been used with success
elsewhere on streets whose traffic is fully equal to the
heaviest in any city of this size. In some few cases
where the grade is more than 1 in 20, or 5 per cent, and
is too steep for asphalt or brick, the use of block-stone may
be justified where they would otherwise be preferable.

COST OF BLOCK-STONE PAVEMENTS.

In the 32 cities reported upon, the cost varies from
the minimum of $1.49¾ per square yard for granite on
sand at Atlanta, Ga., and $1.67 on concrete at Min-
neapolis, to $1.90 per square yard on gravel at Roch-
ester, $2.00 at Portland, Maine, and at San Francisco,
and $2.05 and $2.08 respectively at St. Paul, Minn., and
Wilmington, Del. From these lower rates, the costs
vary up to maximum rates of $3.00 on concrete at
Rochester, N. Y., $4.20 on sand at Cincinnati, Ohio, and
$4.25 on concrete at Detroit, and $4.75 at New Orleans.
The average cost for the 32 cities (most of which use a
sand foundation) is $2.97 per square yard. This does
not include grading and curbing.

SIZES OF BLOCKS.

The standard sizes of block-stone for pavements vary somewhat. The Boston granite blocks are dressed to $3\frac{1}{2}$ by $4\frac{1}{2}$ by 7 inches deep; the New York Belgian blocks of granite or trap are 6 to 8 by 5 to 6 by 6 to 7 inches deep. The more recent form of Guidet pavement used on Broadway consists of granite blocks laid on concrete, the sizes being $3\frac{1}{2}$ to $4\frac{1}{2}$ by 10 to 14 by $7\frac{1}{2}$ to $8\frac{1}{2}$ inches deep. The Medina sandstone blocks used in Rochester are 3 to $5\frac{1}{2}$ by 9 to 12 by 7 inches deep. On Main street these are laid upon six inches of concrete covered with two inches of fine sand; the joints between blocks being filled with hot coarse sand, followed by hot paving pitch until full. This costs, complete, $3.00 per square yard.

JOINTS.

The evils of block-stone pavements are minimized by filling the joints between the blocks with cement. The cement used may be either Portland hydraulic cement or bituminous paving cement; the latter having the peculiar advantage that when a joint has been broken by settlement, its bond is renewed when the cement is softened by the sun's heat. The method of its use is described above.

When the joints are properly filled, surface water is prevented from undermining the blocks and wetting the earth roadbed, as it does when the joints are only filled with sand. When cemented, the blocks support each other, so that their tops are not so soon rounded by the breaking of the edges. There is thus less lodgement of dirt and dust and less noise than when the joints are open.

Chapter V.

Asphalt Pavements.

Comparative Qualities.

Asphalt pavement ranks first in satisfactory qualities, being fairly durable, and clean, and noiseless. In these respects vitrified brick is its only rival for public favor. The first cost of asphalt is generally more than that of brick, but less than the cost of granite blocks.

History of Asphalt Pavements.

The original asphalt pavements were made in Paris in 1854, and were formed of pulverized natural asphalt rock, mined at different places in France and Switzerland and Sicily. This rock is a natural combination of 88 per cent of amorphous carbonate of lime with 12 per cent of mineral tar or bitumen, forming a bituminous limestone. This is now generally used in the European cities, and a similar rock is mined and used on the Pacific coast; but in about 100 cities of the United States an artificial combination, forming a bituminous sandstone, makes a pavement which is as good and in some respects is better, just as artificially combined Portland cement, when proportioned and made with the proper skill and care, may and generally does excel the chance combinations of nature which form natural hydraulic cements.

The great advantage of the artificial mixture over the natural rock asphalt is that the former is less slippery, the sand in the mixture giving a good foothold, so that fewer horses slip upon it and still fewer fall down.

This artificial combination (of which the approved proportions in 1893 are 3 to 5 per cent of powdered carbonate of lime, or limestone dust, 79 to 84 per cent of sand, and 13 to 16 per cent of Lake Trinidad asphalt and petroleum) was first used at Newark, N. J., in 1870, and on Fifth avenue in New York in 1873, though its first extensive use was in Washington in 1877. The proportions are varied, by the experienced judgment of the builder, to suit the local conditions. Since 1883 Buffalo has paved with sheet asphalt 150 miles of streets, having an average width of roadway of 30 feet. This has cost Buffalo over eight million dollars, and equals the combined mileage of all the European cities; London having 26 miles, Berlin 83 miles, Paris 24 miles, and other cities 18 miles, or 151 miles in all.

VARIOUS COMPANIES.

Since 1877 many different methods of construction have been tried, and a number of widely advertised companies are now before the public as builders of asphalt pavements. To be assured of the best results, no offers should be entertained except from some one of the few great firms having an extended experience and an established reputation, who will assume all responsibility for materials and methods, and can give an effective guarantee for a period of ten years; five years does not cover the critical time. Lake asphalt is used exclusively by all these large and experienced firms, to whom only will the company which controls the supply sell the lake asphalt. Smaller local concerns cannot purchase lake asphalt, and are therefore limited to the use of "land

asphalt," or "overflow asphalt," which may or may not
be as good, and always has the fault of uncertainty, and
generally other defects as well. Bermudez asphalt, which
has recently been introduced from Venezuela, is claimed
to be fully equal to that from Lake Trinidad, but the
comparative merits can only be known after years of use.
The details of the materials and construction are
omitted here, but they should be based upon the specifications used in Washington in 1893 by the engineer
commissioners of the District of Columbia. These
specifications are the result of the cumulative knowledge
derived from some unsatisfactory work in the past, and
from a great amount of more recent good work, of which
the general methods have been as follows:

In 1885 the standard pavement consisted of a foundation of six inches of hydraulic cement concrete, covered
with a half-inch "cushion coat" of asphaltic mixture, with
a surface coat of a slightly different proportion, compacted
by rolling to two and one-half inches in thickness.

IMPROVED METHOD OF 1893.

In 1890 a foundation of four inches of bituminous
concrete was used in some cases instead of the hydraulic
cement concrete, as being 25 cents per square yard
cheaper, but in 1893 this was decided to be inferior to
the concrete of hydraulic cement. The detail of the
asphaltic covering, too, has been changed by substituting
a "binder coat" of one and one-half inches finished
thickness for the "cushion coat." This "binder coat" is
a bituminous concrete formed of small crushed rock, combined with one-eighth its bulk of asphaltic mixture. It
is rolled hot till solid, and then covered with the surface
coat of one and one-half inches finished thickness. The

concrete foundation for this pavement is formed over a perfectly sub-drained and solidly rolled dry earth road-bed, formed with a convexity of about one-sixtieth to one-eightieth its width. If too flat, water stands upon the surface and lessens its durability. The cost in Washington City is $2.25 per square yard, not including curbing. In the Northern States asphalt should be laid only during the summer months, as cold weather increases the difficulty of doing good work.

STONE PAVEMENT AS FOUNDATION.

In some of Oswego's business streets, where there are no street-car tracks, the present cobble-stone or block sandstone pavements could be left undisturbed to take the place of the ordinary concrete foundation in places where the convexity of the old pavement is not too great for asphalt. Where the convexity is too great it can be corrected by a strip of filling along the sides next the curbs. Where car tracks exist the company can be required to raise them two inches to allow for the thickness of the asphalt sheet.

The hollows in these stone pavements could be leveled and the spaces between the rounded tops of the stones could be filled with the bituminous concrete forming the "binder coat," and the asphalt surface coat then spread over all. Cobble-stone and block-stone pavements have been thus utilized in New York, Brooklyn, Washington, and elsewhere with an average saving of about 60 cents per square yard. The greater part of the asphalt pavements in New York are laid upon old block-stone pavements as a foundation. One company has, in different cities, laid sheet asphalt upon 18 miles of old macadam road as a foundation with good results.

In several cities where brick pavements have worn badly it has been proposed to use the brick as foundation for a covering sheet of asphalt. It is reported that this has been done on a large scale in Columbus, Ohio, where much brick pavement was built about 1887, which has given poor results.

CAUSES OF FAILURES OF SHEET ASPHALT.

A reasonable amount of traffic tends to prolong the life of a good sheet asphalt pavement. When a properly constructed pavement begins to fail, the causes are probably to be found in about the following order:

First. Defective foundation, which has settled and caused hollows in which pools of water have stood upon the surface of the asphalt until it has become disintegrated.

Second. Patches where the pavement has been torn up for sewer and water connections.

Third. Surface cracks, which sometimes appear in cold weather as a result of excessive contraction of the surface, and which close and reunite in warm weather under the combined effects of warmth and of passing wheels.

Fourth. Excessive traffic, which has worn off the surface. This is least common.

Fifth. Lack of traffic, allowing the asphalt to become spongy.

The latter cause usually shows its effects at the sides of the roadway next the curbs, where there is least passage of wheels. The process of failure may then be as follows:

The material composing the sheet of asphalt expands slightly with the sun's heat, as all other substances do;

but, unlike most other substances, it does not of itself at once return to its original thickness when the heat is lost, because the asphalt becomes rigid as it cools, and unless compressed by force, tends to remain in its expanded form.

In the center of the roadway, where most of the wheels pass, the asphalt is at once re-compressed, but at the sides this is not done so promptly, with the result that there is a tendency to become somewhat porous or spongy where there is little traffic.

When at last the asphalt has actually become porous, water can penetrate it, and this soakage of water is helped by the fact that the surface drainage is toward the sides, where the material is most likely to absorb some of it. Having thus absorbed ever so little moisture, of course both heat and frost have increased effects upon the material, and ultimately it shows signs of disintegration.

All these failures may be provided for in advance, at a known cost, by employing a responsible company who will avoid some of them and who will also guarantee not only to maintain the pavement for ten years, but to deliver it at the end of that time in perfect order; the city thus assuming no risk.

There have been instances where no repairs whatever have been required during or at the end of the guarantee period, but in many other cases the repairs have been extensive.

COST IN VARIOUS CITIES.

The records and reports in 27 cities of the United States show a range in cost per square yard for completed sheet asphalt pavement from minima of $1.95½ at

Utica and $2.00 at Rochester, N. Y., to $3.50 at Buffalo and at Boston; the average cost for 27 cities being $2.81.

These prices include concrete foundation in every case, and also include guarantees for all repairs for periods varying from five to twelve years, and also generally include resetting the old curbing.

The common guarantee period is five years, and when a longer term is agreed upon it is generally upon the basis of an additional payment of about eight cents per square yard per annum, which is a profitable rate for the experienced asphalt companies.

If new curbing is needed it is usually furnished by the city, and its setting costs from 8 to 12 cents per lineal foot of curb.

As asphalt is usually put upon an already improved street, grading is not needed in most cases, but if required it costs an average of 30 cents per cubic yard of material handled.

The rate of $1.95½ at Utica is the lowest recorded, the contract having just been made after peculiarly close competition, which reduced the price 15 per cent.

Nearly as low a rate might be made at Oswego if the present stone pavement served for foundation as suggested, and at such a rate—which would be little over three-fourths what the sandstone block pavements have cost—Oswego could well afford to thus pave some of its business streets; for instance, on East First street past the new opera house, where a noiseless pavement would be most appropriate, and on West First street north of Bridge street, where also there are no car tracks. With proper care and with the construction described in Chapter II, it can be laid where there are car tracks if the latter are raised two inches.

ASPHALT BLOCKS.

Asphalt block pavements are largely used, and cost about 90 per cent of the cost of sheet asphalt. In 1892 the total mileage in the United States was one-eighth that of sheet asphalt. The blocks are usually 4x5x12 inches, are formed of 8 to 12 per cent of asphaltic cement and 88 to 92 per cent of crushed limestone or bluestone, and are bedded in sand upon a gravel or hydraulic concrete base, with their joints filled with dry sand.

Some block asphalt has been used in Syracuse with most unsatisfactory results, but there is no reason why it should not be durable and good.

Chapter VI.

BROKEN STONE ROADS.

TYPES.

In the recent wide discussion of "Good Roads," macadamizing, or some more or less similar arrangement of broken stone, is most often spoken of, and the general reader who has given no special attention to the subject is most likely to conclude that some such construction suits all conditions and localities, though it is really best suited and most used for roads outside of cities.

The two principal types of broken stone roads are the Macadam and the Telford.

In each of these styles of construction, the Macadam material, or road metal, which forms the wearing surface, must have the following characteristics to give good results.

CHARACTER OF MATERIAL.

The rock from which the fragments are to be broken or crushed must be of the toughest and hardest character: first, that it shall be durable: second, that it shall not form dust and mud. For these qualities the different rocks are ranked as follows, in about the order named:

Dolorite, commonly known as trap rock or basalt, is generally considered to be the best. It has no true cleavage and breaks irregularly, and is tough and does not easily grind into dust and mud. The nearest accessible trap formation is the Palisades of the Hudson, in Rockland County, N. Y., about 360 miles by canal and river from Oswego.

Porphery is ranked next, but none occurs nearer than Lake Champlain. *Quartzite* and *quartz* are next, but no definite supply is available. *Gneiss* is next, and this or some closely similar formation exists and is largely prepared and used for roads at Little Falls, on the Mohawk River, 117 miles by canal from Oswego. It is there crushed and shipped by rail and called trap rock; but, as shown by the varied specimens just received, some of it is gneiss, some syenite, and some granite, but none is as tough as trap, though it is a fair road material.

Blast-furnace slag is next, but it is not to be had nearer than at Charlotte, 70 miles west on the R. W. & O. Railroad. Nearly all of the supply there has been engaged for the Rochester boulevards. *Syenite,* somewhat resembling granite, ranks next, but no supplies are near. *Granite* of some varieties ranks high, but much of it crumbles easily and wears into sand. There is none available except water-worn field-stone in small amounts.

Limestone, if unusually hard, is next, but it is only suitable for light traffic, as it crushes, under heavy loads, into fine dust-like clay, which forms a mortar-like mud peculiarly injurious to clothing. Its cementing action, so called, is purely imaginary, and is merely that which an admixture of clay would give to any stone.

The Chaumont limestone quarries, 45 miles distant down the lake, are accessible by sailing vessels or barges, and offer an unlimited supply of quarry-waste which can be had for the cost of transportation.

Sandstone is wholly unfit for any but country roads of light traffic, grinding quickly into dust and mud, so that it is poor economy to use it anywhere. This, of the variety known as grey sandstone, is the only rock of which Oswego has a supply.

IMPORTANCE OF UNIFORMITY.

In the selection of material to be crushed for road metal, uniformity in character is of the first importance.

Material which is uniformly of a *second* grade would be preferable to a mixture of better and worse. Such a mixture of fragments of hard and soft rocks results in quickly crushing the softer pieces, and then exposing the harder pieces to excessive shocks from passing wheels.

FIELD-STONE ARE WORTHLESS.

Rounded cobble-stones, gathered from the fields and lake shore, make the poorest possible road metal, whatever their composition.

Being water-worn into rounded forms, all the fragments crushed from them have at least one curved or water-worn face. These curved and polished faces prevent the adjacent fragments from ever coming to a solid bearing in a roadbed. They will always rock or slide under passing loads, and will loosen all the fragments which touch them. A sketch to illustrate this has been made from actual fragments as crushed from a rounded field-stone and as placed in a road.

Further, these field-stone, which were strewed broadcast over all this region during the glacial period, came of course from the most widely different localities in the northern part of this continent, and include all possible varieties and degrees of hardness.

Field-Stone Fragments.
(Actual)

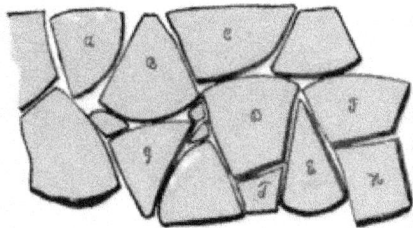

Quarry-Rock Fragments.
(Actual)

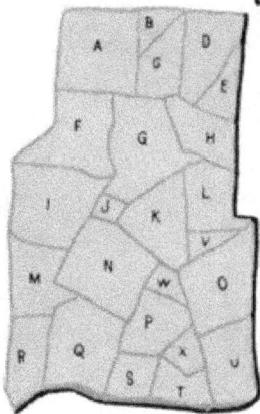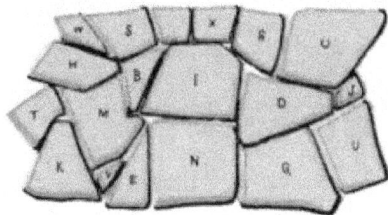

Granite, syenite, quartz, limestone, flint, and slate make up perhaps one-tenth of a mass of them, and the remaining nine-tenths are sandstone, of which at least one-half are so disintegrated by exposure to weather— so "rotten" as the workmen call them—as to be worthless for any purpose. For road surface metal they are worse than worthless, as their only effect is to destroy the good material with which they chance to be mixed.

There is now on hand at the city stone-crusher a large stock of such field-stone ready to be crushed which have cost $2.50 per cord, or 53 cents per cubic yard. No one can afford to use field-stone for road surface material if it were furnished free.

THE MACADAM AND THE TELFORD SYSTEMS.

About a century ago Macadam preached and practiced a gospel of good roads for England with an effectiveness which our league of to-day can only hope to imitate in the United States.

England had long had roads of broken stone, and the use of this material was not peculiar to Macadam's method; but he was the first to establish rules of construction which were generally accepted, and under them were built 25,000 miles of road which formed a network all over England; so that his name has come to be associated with broken stone as a road material, although Telford, who came 25 years later, used the same material but in a different and much more effective manner. In Macadam's talks to committees of Parliament and to his workmen, he always enforced the idea that the whole secret of making a good road was to keep its earth-bed dry; that the ground was the real road and must bear the weight of the stones, as well as of the traffic, and that the subsoil, however bad, would carry any weight if made dry by drainage and kept dry by an impervious covering.

In this requirement Telford and all skillful road-makers fully agree.

This dry roadbed Macadam covered with a layer of road metal of a finished thickness of five to ten inches (varying with the weight of traffic), composed of small, angular fragments of the hardest and toughest rock, broken to a uniform size, as nearly as possible to one and one-half inch cubes, or six ounces each in weight. No dimension larger than two inches was allowed, and any piece too large for a workman to put in his mouth was to be broken again.

NEED OF BINDER WITH BROKEN STONE.

Macadam required that this layer of regular fragments should be spread on the earth roadbed, to be consolidated by the wheels of passing vehicles, without the aid of any fine material or of "binder" of any sort.

This requirement was impracticable and probably could not be enforced, and experience has shown that it is not desirable that it should be enforced.

Such fragments, loosely piled or spread, have about 45 per cent of void spaces, and will pack by rolling to about two-thirds their thickness when loose.

The consolidation of perfectly clean, regular, angular fragments of trap rock, free from screenings or binder of any sort, was thoroughly tried by Mr. Grant in Central Park, New York City, in 1860. A piece of road covered with Macadam's ideal road metal, free from binder, was rolled for several days, until the fragments were worn and rounded, without firm consolidation being effected, and this experience has been repeated elsewhere. Road material which can be thus packed must be of a poor quality, which will supply itself with binder by readily grinding into dust and small pieces. Telford's system differed radically in that he first covered the earth roadbed with a rough pavement of firmly set stones, and that the wearing layer of broken fragments varied in size, and that a binder of fine material was spread over the surface to help in its consolidation.

MODES OF USE OF BINDER.

In England there is now a great variety of methods of construction of broken-stone roads, but as a general thing Macadam's method of using perfectly clean fragments of broken rock is not now followed. The commonest practice seems to be to use 25 per cent of binder, called "hoggin," consisting of loam with coarse sand and gravel, similar to the deposits of bank gravel and loam found near Oswego on the Macfarlane farm and on the Bishop farm and elsewhere. On the English roads, this "hoggin" is washed into the layer of broken stone by flooding the roadway with water.

In France, where the greatest care is given to road construction and maintenance, 25 per cent of sand is generally used with the broken rock as a binder. This is washed in to fill the voids between the fragments of rock, with a final addition of chalky dirt and water to fill the voids in the sand.

In the United States experience has satisfied most American engineers that the roads wear better and have less dust and loose stones if 33 per cent of binder is put on the layers of stone to fill the spaces. The dust and pieces of trap rock which pass through a one-inch screen at the crusher are used for this purpose, or where a softer rock than trap rock must be used, as at Oswego, clean coarse sand may be combined with the screenings, or substituted for them, as making less dust.

Engineers are generally agreed as to the desirability of using a binder for broken-stone road material, but in the matter of Telford's foundation and Macadam's omission of it, there are wide differences of opinion; French and English engineers generally omitting the Telford foundation, and American engineers generally favoring its use.

At Oswego, however, there is little room for discussion on this point, as the entire absence of good rock to be crushed for road material compels that it be brought from a distance, and that a construction be used which requires the least of it. This is of course the Telford, for which the local sandstone will form the foundation course.

CONSTRUCTION OF TELFORD ROAD.

To build a Telford road the earth roadbed must first be sub-drained, and then be cleared of all soft and loose earth and ashes, and the hard earth formed with a regular grade and with a convexity of about one-fortieth the width; that is, a 32-foot roadway should rise in the center about nine and one-half inches after it has settled, for which three inches more should be allowed.

A heavy roller of ten or more tons passing over the earth roadbed will disclose the existence of a surprising number of yielding places and soft spots which could never be found in any other way, but which can readily

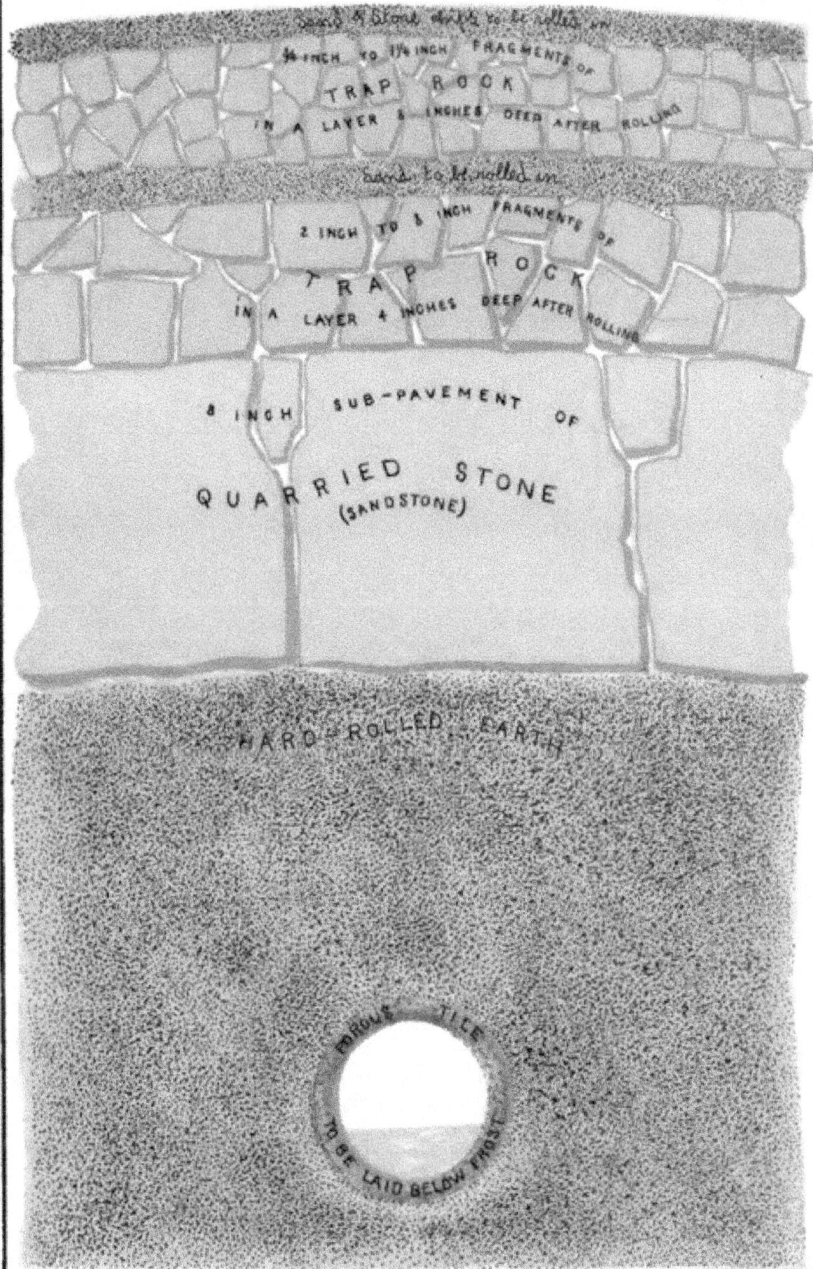

Telford.

Sand & Stone chips to be rolled in

¾ INCH TO 1¼ INCH FRAGMENTS OF

TRAP ROCK

IN A LAYER 6 INCHES DEEP AFTER ROLLING

Sand to be rolled in

2 INCH TO 3 INCH FRAGMENTS OF

TRAP ROCK

IN A LAYER 4 INCHES DEEP AFTER ROLLING

8 INCH SUB-PAVEMENT OF

QUARRIED STONE

(SANDSTONE)

HARD-ROLLED EARTH

POROUS TILE

TO BE LAID BELOW FROST

be filled and rerolled until the earthbed is regular and equally hard throughout.

FOUNDATION.

It is then ready for the foundation or sub-pavement, the stones for which may be of sandstone or of any species of angular quarry rock, in pieces not larger or smaller than 3 to 5 inches thick, 9 to 15 inches long, and about 8 inches deep.

These are set on edge close together, with the longest dimension running across the street, and in lines whose stones break joints with the stones of the adjoining lines; the whole, when set, making a rough sub-pavement eight inches thick. All projecting points are then broken off, and small fragments of stone are then driven into all the irregular cavities between the stones until they are wedged close and solid, so that a heavily loaded wagon can be drawn over them without displacing the pieces. A good workman will average about twenty minutes per square yard in setting this foundation.

CRUSHED ROCK.

Upon this sub-pavement the first layer of crushed trap rock is then spread six inches thick, its fragments varying in size up to three inches, which they must not exceed. The heavy roller will compact this six-inch mass to four inches, and then an inch or more of sharp, clean sand, or waste from the stone-crusher, should be spread over it and rolled in.

Over this a second layer of crushed trap is spread four and one-half inches deep. This is of fragments not smaller than five-eighths of an inch cubes, nor larger than one and one-half inch cubes, but generally about one and one-fourth inch. This is also rolled after wet-

ting until packed to three inches, an inch of screenings
from the crusher or of sharp sand being added to fill the
voids.

SURFACE.

Over this is put a final coat, about three-fourths of an
inch thick, of chips from the crusher or of coarse sand
and fine gravel. This is sprinkled freely and is rolled
until hard and solid.

Sometimes it is required that the rolling shall continue until the material is packed so firmly that an inch
cube of stone laid upon the finished surface shall crush
under the roller without sinking into the road surface.

COST.

The cost of such a roadway in Oswego would probably be from $1.20 to $1.30 per square yard. This
equals $4.26 to $4.60 per lineal foot of 32-foot roadway.

With about this method—except that the three
layers of binder were generally put on top the last layer
instead of between the layers—some of the finest roads
in the United States have been built in and near New
York City by Mr. E. P. North, M. Am. Soc. C. E., and
others, and in New Jersey by Mr. Chas. B. Brush, M.
Am. Soc. C. E., and others, at costs varying from 80
cents and 90 cents to $1.05 per square yard complete.

ROLLING.

If the rolling is done with a 15-ton steam roller, it
would operate at a speed of 1¼ to 1½ miles per hour
for 24 to 26 hours per 1,000 square yards, or 290 lineal
feet of 32-foot roadway, passing over the surface between
120 and 140 times.

MAINTENANCE.

The expense of building a broken-stone road—whether it be Macadam or Telford—is the smallest part of its final cost.

The system of roads which was built early in this century all over England, required then, and still continues to require, the constant attention of an army of resident workmen living along the line of the roads and making never-ceasing repair of ruts and breaks as soon as they occur. Little piles of broken stone, or of stone to be broken, were and are never-absent evidences of constant care, and steam road rollers are often met when driving through the country. Such care is necessary and is very costly.

In London and in Paris broken-stone roads are the roads of luxury; some of the finest streets having macadamized central driveways, bordered on each side by 13 feet of sheet asphalt.

In Paris the annual cost of maintenance of suburban macadamized streets having light traffic is about one-third the original cost of building them. In some cases of extra heavy city traffic, the annual care costs one-third more than the original building; that is, the roadway 14 inches thick has to be practically renewed every nine months. In such cases the so-called cheap macadam is more costly than asphalt, which is therefore replacing it.

In any case, eternal vigilance and a continuing supply of money are the price of a good system of macadam city roads.

This opinion is sustained by the recent reports of the engineer-in-chief of the French system of highways; by

a report of the engineer officer in charge of the highways of the District of Columbia, and more recently by
the author of the paper to which was awarded the first
prize by the League for Good Roads.

Another recent writer who advocates a general use of
macadam roadways says that they can be kept in repair at little expense, not exceeding $10 per mile per
year. He adds on the next page that "such roads
should have constant supervision and attention every day
of the year," but fails to explain how this can be done
for the $10 per mile per year before mentioned.

There is, therefore, no real difference of opinion on
this point between those who favor and those who
oppose the general use of macadam roads for light city
traffic.

Toronto spent $10,000,000 in macadamizing streets
which became seas of mud after every rain, and many
of which were finally dug up to make place for other
roadways better suited to city streets.

This type of roads is generally used for and is best
suited to comparatively long lines of semi-rural roads
or driveways, not thickly built upon nor having very
frequent branch roads.

For use in the streets of Oswego, this suggests an
adverse feature which does not enter into the case of a
rural road. Any one of the principal streets of Oswego
which may be macadamized is crossed every 200 feet
by the existing dirt roads which must for many years
continue to be the character of roadway in the side or
branch streets.

Thus, at intervals of 200 feet, wagons coming constantly from these side streets onto the macadamized

street will bring their wheels full of mud, which will quickly fall onto the road surface. Mud is the deadly enemy of broken-stone roadways, working in between the small angular fragments of stone which have been made and placed with so much care, crowding them apart and destroying their firmness and mutual support.

The durability of the road surface and its resistance to the action of weather and of wear will be in direct proportion to the attention given to removing mud and detritus.

Neglected, the stone fragments of the road-covering will soon be separated from each other by mud, which will finally form half of the mass, which will then be easily cut through by narrow-tired wheels.

It is thus evident that to secure a reasonably good result, not only the main street but all its branches should be macadamized at about the same time. This seems hardly practicable.

The principal objections to this class of roadway for city streets are that their cost is excessive and that they will be muddy during rains and dusty soon afterward, and that they will demand more care and attention than anyone will be willing or likely to give. Those who build will be blamed if those who come after fail to repair. For residence streets, better temporary roadways can be made at one-fourth the cost with clay and gravel.

Chapter VII.

CLAY-GRAVEL ROADS.

USE IN THIS COUNTRY AND ELSEWHERE.

In some parts of the country there are natural deposits of what is usually known as cement-gravel, which is in much favor for road-making. There is really no cement about it, but merely a chance admixture of clay with gravel in proper proportion to make, when wetted and packed, a sort of natural concrete.

The resulting road surface is very smooth, and has a peculiar elasticity which is said to give to those who ride over it the idea of rolling on velvet. It has much less mud and dust than a macadamized road, is easily repaired, and should be comparatively cheap.

At New Orleans such gravel, there known as Rosetta cement-gravel, is brought by the Mississippi Valley railroad 300 miles, and has been used within the past five years to make many miles of roadway in all parts of the city. Captain Dan C. Kingman, of the United States Corps of Engineers, states that at Baton Rouge, La., similar roads are made by an artificial mixture of clay and gravel from the river banks.

In Norway one of the finest highways in the world, leading 150 miles across the mountains to Christiana, is surfaced with cement-gravel and is said to be peculiarly pleasant to ride over.

Nearer home, the new boulevard from Rochester to Lake Ontario along the east side of the Genesee River is surfaced with clay-gravel, and many of the finest rural driveways near New York are thus formed. There is no reason why an artificial mixture should not be made at Oswego of the clay and gravel which can be had in the vicinity, and the mixture be used for road surfaces in the residence streets.

A sample block of this road surface has been made to illustrate this, using one part of lake-bluff clay mixed with two parts of lake-shore gravel. It was not subjected to nearly the same pressure which a ten-ton roller would give, and this is a larger proportion of clay than would probably be used in actual practice.

SUPPLIES OF GRAVEL.

The best gravel is not easy to find in large quantities, and it may prove on investigation that Chaumont lime-stone quarry-waste, crushed small, will be cheaper than gravel, especially as the city crusher is operated by water-power at the merely nominal cost for power of $5.00 per month.

Bank-gravel can be brought here by canal from points eight to twelve miles up the river, one deposit being near Battle Island and another near Fulton. Lake-shore gravel is generally too uniform in size, coarse and fine not being intermixed to any great extent.

On the lines of railroad there are no large gravel banks nearer than Jamesville, just south of Syracuse, on the Binghamton railroad, or Adams on the Rome, Watertown & Ogdensburg railroad, or Briar Hill on the

Utica & Black River railroad; the latter deposit being the largest and best.

If gravel is used, it must be screened of stones larger than one inch, and of the fine dirt. If limestone is crushed, the fragments should not be larger than one inch cubes, and half at least should be less than one-half inch cubes. Gravel is to be preferred.

The clay to be combined with the gravel or crushed limestone occurs along the river and lake banks and at many places in the vicinity. It is hard and tough in its natural formation, but if broken up with a plow or scraper and left for the action of the sun, rain and frost, will soon become manageable.

CONSTRUCTION.

The proper proportion of clay is a slight excess of the voids in the gravel or crushed limestone. One part of clay to two parts of gravel was the exact proportion used in the sample block. One to two and one-half would probably be better in actual use.

The earth roadbed should first be prepared by sub-drainage, if not naturally drained, and by taking off all mud, ashes and loose earth down to the hard earth, which should be given a regular crown of six to eight inches in 32 feet width, and then solidified by the heavy roller, and by filling the hollows which the roller will develop.

On this roadbed the dry mixture of clay-gravel should be spread about four inches thick, well wetted, and then rolled with the steam roller until the clay comes to the surface. Then two more similar layers, similarly rolled while wet, will make a finished roadway about nine inches thick.

This will use, per square yard of surface, one-fourth cubic yard of gravel or limestone fragments, and one-tenth cubic yard of clay.

COST.

The cost ought not to exceed 30 cents per square yard. With such arrangements for screening and mixing as might be particularized, it could probably be done for 25 cents. This is one-fourth the estimated cost of Telford road.

Chapter VIII.

VITRIFIED BRICK PAVEMENTS.

During the past ten years there has been a steadily increasing use in the United States of vitrified brick for the pavements of the streets of cities and towns of moderate size—that is, of 100,000 inhabitants and less; the larger places welcoming the new material as being a competitor with sheet asphalt and as offering another means of escape from the intolerable noise and dirt resulting from block-stone pavements, while the smaller western towns, with characteristic enterprise, have built miles of brick pavements to displace the natural mud.

The accompanying illustrations showing brick pavement in process of construction and after completion are from a report by Mr. F. A. Dunham, C. E., upon work under his charge.

EXTENT OF ITS USE.

Two to three hundred such cities and towns have laid vitrified brick pavement, and its use is constantly increasing.

The comparative extent of this use of brick is indicated by the replies made in May, 1894, to *Municipal Engineering* as to proposed paving by 75 cities in all parts of the United States. None reported any proposed block-stone work; seven reported proposed asphalt pavements, and 20 reported proposed brick pavements, the

STREET BEFORE PAVING.

total of the amounts to be expended for brick being five times that for asphalt.

These extensions have been most marked in the past two or three years, following the period covered by the very complete articles published in the *Engineering News* in September, 1890, July, 1892, and August, 1893; the first by Charles P. Chase, C. E., the second by Robert Gillham, M. Am. Soc. C. E., and the last by Daniel W. Mead, M. Am. Soc. C. E.

REACTION AGAINST USE OF BRICK.

On the other hand, there has undoubtedly been a reaction in the popular desire for brick pavements in some of the larger cities, where people have learned to know what good pavements are, and where sheet asphalt has been brought into close comparison with brick.

The comparatively noisy brick pavement is objected to on some residence streets, and on streets having heavy traffic there have been some poor results as to durability. Much discredit has also been thrown upon the use of vitrified brick by the careless and ill-judged manner in which many manufacturers have sent out irregularly and imperfectly burned brick. These have been laid and have caused the failure of many pavements, thus stopping further extensions and preventing some other cities from using brick at all, to the great gain of the sheet asphalt companies.

INDICATIONS OF POPULAR OPINION.

A Syracuse street which was paved with Syracuse brick three years ago has now required extensive repairs, and of the new contracts made this year for Syracuse pavements 4,777 square yards, or about one-fourth of a mile, are to be of brick, and 166,000 square yards, or about ten miles, are to be of sheet asphalt.

Watertown has had Syracuse brick pavement for two years, but is now laying sheet asphalt extensions. These facts are in no way conclusive as to relative merits, but are worth considering as indications of popular opinion based on experience.

RECORDS OF ACTUAL RESULTS.

To bring down to date the records of this modern revival of an ancient paving material (long used in Japan and Holland), the city engineers of 64 cities of the United States known to have brick pavements have been written to, and 50 have replied by filling the blanks of the following circular:

....May................1894.

The city of...has about................
miles of streets which have been paved with vitrified brick for
................to.........years, with.........................results.   In our
best work, the top course is laid upon...............................as a
foundation, and its joints are filled with............................ .   The
average cost has been............. ....per square yard.

....
                                             City Engineer.

The records of the others have been filled out from the *Engineering News* articles and elsewhere, and the following table compiled:

SUMMARY OF REPORTS OF MODES OF CONSTRUCTION, COST AND RESULTS OF VITRIFIED BRICK PAVEMENTS.

CITY AND STATE.	Miles in use June, 1894.	Cost per Square Yard of "Best Work" on the Foundation here indicated.			Filling of Joints.	Reported Results.
		Six inches Concrete.	Flat Brick or Gravel.	Broken Stone or Gravel.		
Atlanta, Ga........	1.1	$2.19	....	....	Paving tar.	Satisfactory.
Atchison, Kan.....	2.75	....	$1.75	....	Sand,	Excellent.
Alton, Ill..........	1	....	2.16	....	Sand.	....
Alleghany, Pa.....	2	1.60	....	....	Paving tar.	Fair.
Bellaire, Ohio.....	....	....	....	$0.61	....	....
Binghamton, N. Y.	0.25	2.40	....	....	Cement grout.	Fair.
Bloomington, Ill...	6	....	2.00	....	Sand.	Good.
Buffalo, N. Y.....	3.33	2.75	....	....	Cement grout.	Fair.
Burlington, Ia.....	7.50	....	1.60	....	Sand.	Gratifying.
Cedar Rapids, Ia...	2	....	1.35	....	Sand.	Fair.
Charleston, W. Va.	....	....	....	1.15	Sand.	....
Chicago, Ill........	1	2.30	....	....	Paving tar.	Satisfactory.
Cincinnati, Ohio...	15	2.50	....	....	Paving tar.	Fair.
Clinton, Ia.........	10	....	....	1.45	Sand.	Good.
Columbus, Ohio...	30	2.00	....	....	Paving tar.	Good.
Connellsville, Pa...	2	....	....	2.49	Sand.	Excellent.
Council Bluffs, Ia..	5	....	1.50	....	Sand.	Good.
Davenport, Ia.....	6	....	1.60	....	Sand.	Good.
Dayton, Ohio......	6.4	2.30	....	....	Cement grout.	Good.
Decatur, Ill........	15	....	1.75	....	Sand.	Good.
Detroit, Mich......	9.6	2.50	....	....	Paving tar.	Fair.
Des Moines, Ia.....	10	1.70	....	....	Paving tar.	Good.
Dubuque, Ia.......	1.5	1.69	....	....	Sand.	Satisfactory.
Dunkirk, N. Y.....	2.5	2.10	....	1.87	Cement grout.	Good.
Evansville, Ind....	4.5	1.70	....	....	Sand.	....
Findlay, Ohio......	4	....	....	1.75	Paving tar.	Satisfactory.
Fort Wayne, Ind..	2	1.63	....	....	Cement grout.	Good.
Galesburg, Ill.....	12	....	1.80	....	Sand.	Good.
Hannibal, Mo......	1.5	....	2.05	....	Sand.	Perfly. satisfy.
Hartford, Conn....	0.12	4.00	....	....	Cement grout.	Good.
Indianapolis, Ind..	8.7	2.35	....	....	Paving tar.	Good.
Jacksonville, Ill...	9	....	....	1.40	Sand.	Good.
Kansas City, Mo...	10.25	2.00	....	....	....	Fair.
Kenosha, Wis......	1	....	....	1.55	Sand.	Good.
Keokuk, Ia........	1.25	....	1.55	....	Sand.	Good.
Lafayette, Ind.....	2.50	....	1.80	....	Sand.	Good.
Lancaster, Pa......	0.10	1.80	....	....	....	Good.
Lexington, Ky.....	6	2.25	....	....	Paving tar.	Good.
Lincoln, Neb......	15	....	1.75	....	Cement grout.	Good.
Lockport, N. Y....	10	2.09	....	....	Cement grout.	Excellent.
Louisville, Ky.....	10	1.50	....	....	....	Good.
Massillon, Ohio...	9	....	....	1.40	Sand.	Good.
Memphis, Tenn....	2.25	2.65	....	....	Paving tar.	Entirely satis.
Olean, N. Y.......	1.50	2.00	....	....	Cement grout.	Good
Omaha, Neb......	10.25	1.87	....	....	Sand.	Moder'tely fair
Ottawa, Ill........	2.25	....	1.40	....	Sand.	Good.
Peoria, Ill.........	7	....	1.75	....	Sand.	Fair.
Philadelphia, Pa...	20	2.05	....	....	....	Good.
Providence, R. I...	1	3.00	....	....	Paving tar.	....
Quincy, Ill........	6	....	....	1.80	Sand.	Good.
Rochester, N. Y...	3.14	2.30	....	....	Paving tar.	Good.
Rockford, Ill......	1.82	....	1.75	1.37	Sand.	Good.
Rock Island, Ill....	7	....	1.62	1.33	Sand.	Satisfactory.
St. Paul, Minn.....	0.34	....	2.40	....	Sand.	Indifferent.
Scranton, Pa.......	0.10	2.33	....	....	Cement grout.	Good.
Springfield, Ill....	5.38	....	....	1.35	Sand.	Good.
Steubenville, Ohio.	10	....	....	1.00	Sand.	Good.
Syracuse, N. Y....	5	2.15	....	....	Cement or tar.	Good.
Terre Haute, Ind..	1	....	....	2.25	Cement grout.	....
Toledo, Ohio......	16.33	....	....	1.05	Sand.	Good.
Troy, N. Y........	1	2.50	....	....	Cement grout.	Good.
Washington, D. C..	0.25	2.05	....	....	Cement grout.	Good.
Watertown, N. Y..	0.12	2.46	....	....	Sand.	Good.
Wheeling, W. Va..	2	....	....	1.35	Paving tar.	....
Wilmington, Del...	3	....	....	2.15	Cement grout.	Satisfactory.
Average of prices.		$2.19	$1.75	$1.52		

REGION OF PRODUCTION.

The use of vitrified paving brick has been in a measure restricted to cities and towns located within the two regions of Pennsylvania and Ohio on the southwest, and Indiana and Illinois on the west, which produce the special quality of blocks demanded, which differ entirely from ordinary building bricks in both their material and their mode of manufacture and in their qualities; the name being a misleading one, as they are not brick, but tile, and are not actually "vitrified."

CHARACTERISTICS.

The clay to mould the brick must be of a peculiar character which will not melt and flow when exposed to an intense heat for a number of days, but will gradually fuse and form vitreous combinations throughout while retaining its form.

The resulting brick must be a uniform block of dense texture in which the original stratification and granulation of the clay has been wholly lost by fusion which has stopped just short of melting the clay and forming glass. The clay while fusing must shrink equally throughout, thus causing the brick to be without any laminations or any exterior vitrified crust which differs from the interior. Such a brick will be incapable of absorbing any considerable amount of water, and will have great strength and toughness.

There seems now to be no difficulty with rigid inspection in getting brick which will uniformly possess these qualities, though many which lack them are made and sold, to the manifest injury of the reputation of paving brick. A number of samples of very poor paving brick have recently been sent by different makers to the Board of Public Works of Oswego.

TESTS.

The comparative merits of different bricks are readily determined by tests. But the real value of any particular kind of brick is best learned by the actual results of service in other cities, and no brick should be contracted for until similar ones have been examined in actual use and found to show little wear after several years in a street pavement.

The tests and their relative values are fully discussed in a valuable pamphlet by Mr. M. D. Burke, C. E., recently published in Cincinnati, in which the general subject of "Brick for Street Pavements" is well presented to the Engineer's Club of that city.

TEST OF ABSORPTION.

The simplest test consists in drying the sample bricks thoroughly for a day by uniform heat, weighing accurately on scales reading to one-sixteenth of an ounce, immersing in water for three days, wiping off carefully all free water, weighing again accurately on the same scales and noting the percentage of added weight resulting from the water absorbed. After making a few such tests of any brand of brick, an expert can tell by looking at the size and color of a specimen how much water it will absorb. It is usual to reject those which will absorb over 2 per cent, and this specification is so easy of application that it is the one most commonly made. Twenty-eight specimens thus tested by Mr. Burke varied in absorption from nothing to over 7 per cent, averaging 1 4/10 per cent; other published tests show a similar range in results, and are chiefly valuable for comparison only.

It is generally considered, however, by Mr. Burke and by other civil engineers who have made a special study of the subject, that the absorption test gives a less valuable indication than any of the other tests, and that a vitrified brick may absorb over 2 per cent of water and yet be proved by other tests, and by use, to be a satisfactory paving brick.

It seems to be accepted that the most valuable indications are obtained from tests by abrasion and impact, and by tests of crushing strength and of transverse strength.

TEST BY ABRASION AND IMPACT.

The test by abrasion and impact consists in putting the sample bricks (after weighing them) into a foundry "rattler" or tumbling-barrel, into which are also put several accurately weighed blocks of tough, close-grained granite, which are similar in size and form to the brick, and are to serve as a standard for comparison. There are also put in with the bricks and the granite blocks 100 to 200 pounds weight of cast iron in fragments of various sizes.

The tumbling-barrel is a cylinder mounted upon a shaft, which is then revolved about 30 times per minute for several hours, and the abrasion of the brick and of the pieces of granite determined by reweighing and noting the losses of weight. The loss from the best brick should not be over two to two and one-fourth times the loss from the granite.

A case is reported from New Haven, Conn., in which two bricks showed no appreciable loss after five hours in the rattler, while the 18 pieces of pig iron had lost 1½ per cent. Other reports show somewhat similar results which are less incredible than this. It is evident, however, that the test affords a good, and perhaps the best, means of comparison, as well as giving some idea of the real value for pavement. A rough estimate which is frequently made is that half an hour in the "rattler" equals a year's service in a pavement under heavy traffic, but this cannot be verified.

CRUSHING STRENGTH.

The test of crushing strength and of transverse strength are both most valuable ones, but they require a special testing machine which is not always available, and a degree of care and skill which few can give to the tests or to the study of their indications.

For the crushing tests two-inch cubes must be sawed from the center of each specimen. Bricks not uniformly fused will, of course, fail under this test, and makers of poor brick will strenuously object to it. Good brick will, however, show a strength of 13,000 pounds or more per square inch: that is, a two-inch cube will sustain a pressure of 52,000 pounds, closely approaching the strength of a block of good granite. Ordinary building brick vary in strength from 500 pounds per square inch upward.

TRANSVERSE STRENGTH.

For the test of transverse strength, a full-size brick is placed on edge upon two knife-edge supports six inches apart, and the breaking load is applied at the middle point by another knife-edge, and is increased until the specimen breaks down. Forty-five specimens thus tested by Mr. Burke varied in strength from 4,580 pounds to 15,170 pounds, and led to the conclusion that a satisfactory specimen should bear at least 8,000 pounds thus applied, giving a modulus of rupture of 1,600 pounds; the modulus of rupture being computed by the formula

$$R = \frac{3\,W\,l}{2\,b\,d^2}$$

in which W = the breaking weight in pounds, b, d and l the breadth, depth and length respectively, all in inches, and R the modulus of rupture in pounds.

BRICK AVAILABLE FOR OSWEGO.

Most of the factories of vitrified brick are far removed from Oswego. There is, however, one located at Syracuse, which uses clay taken from the "Old Brickyard Point" on the Oswego River, 25 miles south of Oswego, and whose bricks are highly commended by the engineers of many cities which have used them.

These favorable reports have been only in a measure confirmed by examinations which have been made of several miles of pavements formed of these brick, which were generally found, after one to two years use, to be in only fair condition; the corners and edges being much broken and rounded, apparently by blows from the calks of horseshoes, this force being the most trying which the pavement encounters.

The peculiarly small size of the Syracuse brick is an objection to them, 81 being required to cover a square yard, while 60 to the square yard is not an uncommon size of other bricks; the smaller bricks giving one-third more edges and corners to be chipped and broken. But a good result of this smaller size is an unusually perfect uniformity of fusion or vitrification.

If the Syracuse brick are decided to be satisfactory after further examination, then a paving brick is available for Oswego at a fair rate, as these can be delivered by canal at small cost for transportation. Another brick which is highly commended, and which is now used in Syracuse, is made in Corning, N. Y.; while western brick can, if necessary, be brought to Oswego by lake at low freight rates. There is, therefore, no doubt of a supply.

VARIOUS STYLES OF CONSTRUCTION.

Before estimating upon the cost per square yard for such pavement in Oswego, it will be best to review the reports of different styles of construction used in other cities, as given in the foregoing table, and to decide which mode is the best suited to the local conditions.

ON CONCRETE FOUNDATION.

By examination of the table on page 47 it will be seen that of 62 cities having vitrified brick pavements there are 34 which use a concrete foundation six inches thick, upon which (with a one-inch sand cushion) a single course of brick is set on edge. One-fourth of these report the results as "fair," instead of the usual "good" of the other reports; but this may in part result from a higher standard of comparison, as indicated by the use of the best form of foundation.

COST.

The average cost of this construction complete, not including curbing and extras, is $2.21 per square yard, ranging from $1.56 at Alleghany, Pa., to $3.00 at Providence, R. I.

NEED OF CONCRETE BASE.

For a city which has been educated to a correct idea of what constitutes a good pavement, the extra cost of the concrete base must be incurred in order to give satisfaction. Where the streets are of made ground, or were formerly swampy or unstable, or in any case where traffic is exceptionally heavy, a concrete base is necessary to insure permanence, and is worth the extra cost.

SINGLE COURSE ON GRAVEL.

Of the 39 cities not using a concrete foundation, 19 use a foundation of six inches of sand or broken stone, or both, upon which is a single layer of brick set on edge. This is the cheapest construction, and ranges in cost from 61 cents per square yard in June, 1894, at Bellaire, Ohio, to $2.49 on extra heavy stone and gravel foundation at Connellsville, Pa.; the average cost of 19 cities being $1.51 per square yard. This is not a good construction, the brick usually settling irregularly and being liable to breakage.

TWO COURSES ON BROKEN STONE.

Eighteen of the 62 cities report the use of *two* layers of brick set upon 6 to 12 inches of broken stone and sand. This is the best construction where concrete is not necessary, and is the one shown in the accompanying illustration, which should indicate the joints as being filled with hot sand and hot paving-tar. It is specially adapted to streets whose sub-stratum is hard-pan or glacial drift which is not liable to settle, as is the case on many of the streets of Oswego.

MODE OF CONSTRUCTION.

The earth roadbed being sub-drained and hard-rolled, as before described for other pavements, is covered by a layer of crushed stone fragments, which may be of quarried sandstone, and should be 6 to 12 inches thick after packing by the roller, the thickness varying with

Vitrified Brick.

SAND TO FILL JOINTS

VITRIFIED BRICK
FIRST QUALITY

1 INCH SAND

VITRIFIED BRICK
SECOND QUALITY
(HALF-PRICE)

3 INCHES SAND
OR COAL ASHES

8 INCHES TO 10 INCHES OF
CRUSHED ROCK
(MAY BE SANDSTONE)

HARD ROLLED EARTH

POROUS TILE
LAID BELOW FROST

LAYING VITRIFIED BRICK ON CONCRETE BASE.

the degree of traffic. This should be formed with a regular "crown" of about one one-hundredth the width. Upon this about three inches of clean sand or coal ashes should be uniformly spread and lightly rolled, leaving it three inches thick, and upon this the first course of vitrified brick is laid flat, in lines lengthwise of the street, using care not to tread upon or disturb the regular surface of the sand or ashes. The brick should be so laid in each line as to break joints with those in the adjoining lines. The brick in this first course should be equal in vitrifaction and strength to the best, but may be slightly warped, and hence of second quality, which will make them of half-price.

SETTING AND TAMPING TOP COURSE.

Over these the surface course of brick is set on edge upon an inch of clean sand, the brick being set in regular and true lines, running crosswise of the street and breaking joints with those in the adjoining lines. The top course should be bedded and settled by tamping with a 75-pound rammer striking upon an eight-inch plank until the rammer rebounds. This will leave the brick in vertical positions, while a roller will cause the brick to lean to the direction in which the roller moves.

COST.

The average cost, in 18 cities, of these double brick pavements complete was $1.73 per square yard, ranging from $1.35 at Cedar Rapids, Iowa, to $2.40 at St. Paul, Minn. First quality vitrified brick should be delivered at Oswego for $11.50 per thousand for "firsts," and $5.75 per thousand for "seconds." One hundred lineal feet of 32-foot roadway would take 28,800 "firsts" and

14,400 "seconds" of the smaller size referred to on page
53. The cost in Oswego should be within $1.95 per
square yard, itemized as follows:

Excavation, one-fourth cubic yard at 32 cents............	$ .08
Roller and extra labor.......	.14
One-sixth cubic yard broken sandstone at 84 cents.......	.14
Sand, one-ninth cubic yard at 45 cents.................	.05
Flat brick, 45 at $5.50 per thousand..................	.26
Edge brick, 90 at $11.00 per thousand.................	1.03
Laying brick, 1 35 per square yard..................	.10
Hot sand and paving-tar joints........	.10
Supervision, etc....................................	.05
Total per square yard...........................	$1.95

FILLING OF JOINTS.

Various methods are employed for filling the surface
joints between the brick, the comparative merits being
about in the order named. Hot paving-tar or coal-tar,
followed by hot sand, makes good joints if carefully
poured between the brick and not allowed to spread over
the surface. If it becomes necessary to take up the
pavement thus laid, the paving-tar can be chipped off
on cool days at a cost of about $4.00 per thousand, and
the brick then relaid. Clear Portland cement grout will
make a good joint, but cannot be removed to relay the
brick. Equal parts of Portland cement and sand mixed
and spread dry and then washed into the joints, is a
peculiar method sometimes used. Clear sand is used in
about one-half the cases shown in the table and costs
about five cents per square yard, or one-third to one-half
the cost of the other modes described, any of which are
better than sand alone The brick will be much more
durable if the joints are well cemented.

FILLING JOINTS WITH HYDRAULIC CEMENT GROUT.

STREET AFTER PAVING WITH VITRIFIED BRICK.

AVOIDABLE FAILURES.

Almost every city using brick pavements has had some portion of the work which has been unsatisfactory; but these failures can usually be traced to needless defects which could have been avoided in the choice of materials or in the mode of construction.

ADVANTAGES.

The special advantages which are offered by the use of vitrified brick are:

First. Less first cost than sheet asphalt, which is its only competitor.

Second. Less ultimate cost, as few repairs are needed if good brick are used.

Third. Ease of construction and of repair, little skilled labor being required. An experienced foreman can teach a gang of ordinary laborers in two days. Such men, each with a helper, have laid 15,000 brick each per ten hours of ordinary work.

Fourth. Ease of traction and of cleaning, and freedom from dust and mud.

Chapter IX.

Conclusions.

LOCAL APPLICATION.

The statements which have been presented in the preceding chapters seem to lead to the following conclusions regarding the pavements and roadways which are best suited to the local conditions:

First. For the main business streets having heaviest traffic, which are now paved with cobble-stone and with block sandstone—like West First street and East First and Second streets, and East Bridge street from First to Fourth—sheet asphalt laid upon these old stone pavements as a foundation and guaranteed for 10 years as described on page 23. This for such main streets only as have grades of not more than 3 per cent.

Second. For the main business streets having grades steeper than 3 per cent—such as East Bridge street from the bridge to First street, and East Utica street from the bridge to Second street—block sandstone laid upon a six-inch concrete base, with joints of hot sand and paving-tar like that described on page 19 as in use on Main street in Rochester, N. Y.

Third. For semi-business streets and thoroughfares—like West Second street and West Bridge street from Third street to Eighth street, and East Bridge street from Fourth to Tenth—vitrified brick laid upon a six-inch concrete base as described on page 53, with joints of hot sand and paving-tar as described on page 56.

Fourth. For residence streets upon which there is most driving—like West Third street and West Fifth street, and East Fourth street—sheet asphalt guaranteed for 10 years, as described on pages 21 and 22. In cases where property-owners will not consent on account of the cost, a cheaper pavement, which will give good results, can be had by using the vitrified brick construction described and illustrated on pages 54 and 55, but for West Fifth street asphalt should be preferred.

Fifth. For temporary and immediate improvement of the residence streets generally, beginning with West Fifth street and East Fourth street, the dirt roads should be sub-drained, narrowed, formed and rolled as described on pages 2 to 5. Upon streets where the construction of permanent pavements is improbable for an indefinite period, this temporary treatment should be followed by making a top surface of clay-gravel as described on page 42; this to be maintained as needed until some permanent pavement can be built.

ULTIMATE COST.

No estimate is given of the sum total of the cost of such a system when it shall have been completed, because it is idle to conjecture how much will be done, and there is now no thought of doing anything more than to make a good beginning and to start upon correct lines.

For any approximate computations it may be sufficient to state that a mile of narrowed roadway, 32 feet wide, paved at the rate of $2.50 per square yard, will cost $47,000, and that the business streets, 54 feet wide, will at the same rate cost $80,000 per mile.

The data for any desired estimates have been given, but the present concern is rather to decide how and when the work shall begin than to predict what it shall have cost when posterity has finished.

COMMENTS.

NEW YORK.

From the *Engineering News*, New York, January 17, 1895.

Although this volume was written primarily to further the cause of good roads and pavements in Oswego, nearly all the matter presented is equally applicable to other cities, and much of the information it contains was collected by the author for practical use because he could not find it elsewhere. The different classes of pavements are described in a clear and concise manner, and details of construction and cost are given. To secure the most recent figures possible in relation to brick pavements, Mr. Judson sent circulars to the city engineers of a number of cities in which brick is in use as a paving material. To these inquiries 50 replies were received, which, with figures from articles in *Engineering News* and elsewhere, bringing the number of cities up to 65, are presented in tabular form in the book. This table gives for each city the miles of brick pavements in use in June, 1894, the cost per square yard of "best work" on different foundations, the kind of joint filling used, and the results reported. Of the illustrations three are colored plates showing practical points of construction and four are street scenes reproduced from photographs. The book would be useful for distribution among city officials where educative work is necessary to show how roads should and should not be constructed.

From the *Engineering Magazine*, New York, March, 1895.

Although a great many books have been written upon the subject of pavements and road maintenance, this little book will have a place among the literature of road-making from the fact that it treats of roads specially adapted to towns of about the size of Oswego—that is to say, containing about 25,000 inhabitants. It discusses the various methods which may and will be proposed for such towns, and the reasons why some are to be avoided and some are to be accepted and used. The author thinks that towns of this kind, contemplating the construction of streets, should profit by the costly experiences of others so far as applicable, and thus avoid building apparently cheap roads whose repairs will afterwards make them cost more than those originally the most costly, or else costly ones which are not adapted to local needs or means. The whole discussion is based upon general reports and results culled from statistics of 75 cities, together with the recently published floating literature upon the subject, written by experienced engineers and published in the technical journals. The book cannot fail to be of value to many municipal engineers and bodies of municipal officials.

From the *Engineering News*, New York, April 11, 1895.

This book should, of course, be owned by every city engineer. It will also prove of great interest and value to municipal officials who are interested in road and street construction.

INDIANAPOLIS.

From the *Paving and Municipal Engineering Magazine*, of Indianapolis, April, 1895.

It is more than a consulting engineer's plan for work, as it is intended to influence the forming of public opinion, and is applicable not alone to Oswego, but to other towns of from 20,000 to 30,000 population. In discussing the comparative qualities of pavements he gives asphalt first rank and credits vitrified brick with being "its only rival for public favor" in the qualities which distinguish it—namely, durability, cleanliness, and noiselessness. He describes other paving methods and materials, however, giving such information regarding them as may lead to the formation of judgment as to their merits.

PHILADELPHIA.

From the *Journal of Association of Engineering Societies*, Philadelphia, January, 1895.

While this work is prepared with special reference to the needs of the City of Oswego, New York, it is believed that it will find application in many other cities of this country. * * * The work is very handsomely gotten up, and is profusely illustrated with photographs and with colored lithographs of the different forms of construction. The inadvisability of using broken cobblestones for road material is well illustrated in one of the latter.

BOSTON.

From the *L. A. W. Bulletin and Good Roads*, of Boston, April 12, 1895.

CITY ROADS AND PAVEMENTS is a 60-page book just issued by William Pierson Judson, of Oswego, N. Y. In the preface he says: "The present purpose is not the making of pavement, but the forming of public opinion." He recommends, among other good things, that cities employ no wagons except those having wide tires and varying track. He shows by drawings the advantages of quarry rock for the production of crushed stone and the objection to round field-stone on account of the curved, smooth surfaces which tend to work upon each other. Colored plates show the more approved methods of laying Telford road and brick pavement, and also of underdraining roads and streets. The book is very interesting, and shows much thought and research on the part of the author. Although Mr. Judson has prepared this work at the request of Mayor Higgins, of Oswego, and with especial reference to that city, it is worthy of a much wider circulation, and would be very interesting to men in charge of the streets of any city.

WASHINGTON, D. C.

From the United States Department of Agriculture.

OFFICE OF ROAD INQUIRY, GEN. ROY STONE, Engineer,
WASHINGTON, D. C., April 2, 1895.

I am much pleased with your book on city roads and pavements, and am recommending it wherever we have inquiry for information relating to that branch of highway construction.

OSWEGO.

From the *Daily Times*, Oswego, April 20, 1895.

The *Times* has said that Oswego was peculiarly fortunate in having her local needs and conditions so ably discussed by a gentleman so deeply interested and wholly familiar with them, making them the basis of a book which had been so widely commended. That a local book should be noticed at all by the great engineering papers and magazines of New York, Boston, Philadelphia, Indianapolis and Washington, was not to be expected; but that it should be so strongly endorsed by them, is evidence that its statements and deductions may be generally accepted and used.

From the *Daily Palladium*, Oswego, April 20, 1895.

When Mr. Judson's book on "City Roads and Pavements suited to Oswego" was issued, the *Palladium* reviewed it carefully and pronounced it to be "easily the most interesting and valuable presentation of the subject of street paving that we have ever seen." The book has now been before the public for three months, and has had a wide circulation. It has attracted much attention among the engineering magazines and papers which make the subject a specialty, and they have reviewed it at length. These unexpected and favorable comments, coming as they do from such papers as the *Journal of Association of Engineering Societies*, Philadelphia; the *Engineering News*, New York; the *Engineering Magazine*, New York; the *Paving and Municipal Engineering Magazine*, Indianapolis, and the *L. A. W. Bulletin and Good Roads*, Boston, go to confirm the *Palladium's* original opinion, and show that the officials of Oswego, or of any similar city, can safely be guided by the general lines laid down in this book, which the *Engineering News* puts forward as being the paving book of the year, and one which will be of permanent value to municipal officials.

www.ingramcontent.com/pod-product-compliance
Lightning Source LLC
Chambersburg PA
CBHW021954190326
41519CB00009B/1253